Management Division

Held at the 2023 AIChE Annual Meeting

Orlando, Florida, USA
5-10 November 2023

ISBN: 978-1-7138-9307-3

Printed from e-media with permission by:

Curran Associates, Inc.
57 Morehouse Lane
Red Hook, NY 12571

Some format issues inherent in the e-media version may also appear in this print version.

Printed with permission by Curran Associates, Inc. (2024)

For permission requests, please contact AIChE
at the address below.

AIChE
120 Wall Street, FL 23
New York, NY 10005-4020

Phone: (800) 242-4363
Fax: (203) 775-5177

www.aiche.org

Additional copies of this publication are available from:

Curran Associates, Inc.
57 Morehouse Lane
Red Hook, NY 12571 USA
Phone: 845-758-0400
Fax: 845-758-2633
Email: curran@proceedings.com
Web: www.proceedings.com

TABLE OF CONTENTS

LEADERSHIP AND MANAGEMENT OF SUSTAINABLE SOLUTIONS IN INDUSTRY

LEADING AND MANAGING INDUSTRY-ACADEMIA COLLABORATIONS AND PARTNERSHIPS

NON-TRADITIONAL CAREER OPPORTUNITIES IN CHEMICAL ENGINEERING FOR MBA'S, LEGAL PRACTITIONERS, AND PH.D.'S

380a Presentation of 2023 Young Leader Award to Dr. Ran Xu of Dupont

Mark Swientoniewski

Mark Swientoniewski, Chair of the Management Division's Awards Committee, will present the 2023 Young Leader Award to Ran Xu of the DuPont Company. The award which is sponsored by the Chemours Company recognizes a younger chemical for demonstrating outstanding achievements in leadership.

380b Putting Inclusive Culture and Core Values at the Heart of Innovation, Sustainability and Digital

Ran Xu

Market-back and technology-driven innovation with positive impacts for our customers and society, a sustainability lens that significantly broadens value creation options, and a strategy enabled by digital allows DuPont to fulfill its purpose to empower the world with essential innovations to thrive. This talk summarizes YLA award recipient Ran Xu's career journey on innovation, sustainability and digital while putting an inclusive culture and DuPont's core values at the heart of this trifecta at DuPont's Centre of Excellence in Process Scale-up.

380c Leading Sustainability in Dynamic Times to Have a Positive Impact, Create Value and Engage Customers, Stakeholders and Employees

Scott Collick

A chemical engineer's perspective on the sustainability challenges facing the planet today and the important role of product and process innovation. Sustainability has become a critical issue with significantly increased regulatory and stakeholder expectations. This presentation will focus on addressing the world's most pressing sustainability challenges (including climate change, circularity, nature, water, and substances of concern) using science and innovation and how DuPont is working on lowering its footprint and expanding its handprint through its 2030 sustainability goals.

380d Challenges in Scaling Biomass-Based Chemical Processes

Keith Hutchenson

Covation Biomaterials is focused on innovation to produce and supply sustainable, high-performance, biobased materials and has a mission of sustainability at scale. Process development for biobased materials involves numerous challenges which are unique relative to the petro-based chemical industry. This talk will focus on some of these challenges and provide examples of the use of biomass for the production of value-added chemicals.

284a Presentation of 2023 Management Award to Shawn Feist of Dow Chemical

Mark Swientoniewski

Mark Swientoniewski, Chair of the Management Division Award's Committee, will present the 2023 Management Award to Shawn Feist of Dow Chemical. The Award is sponsored by DuPont de Nemours, Inc and recognizes a chemical engineer for outstanding achievements in the management and leadership of chemical engineers.

284b Beyond the Technology: All the Stuff I Wish I Would Have Known Sooner

Shawn D. Feist

As one often reflects on a career, it is common to make the statement, "I wish I would have known that sooner." As scientists and engineers a great deal of effort is put in place to be excellent technically, for obvious and necessary reasons. Yet often when a "failure", or better put a "learning" occurs, it frequently is not the technology that failed. As a leader it is critical to navigate yourself and your team through these "failures" and persevere and learn from them. The first important step in that process of learning is to spend time introspecting to diagnose the failure and then be humble enough to admit the root cause, which is often not the technology in which we have spent years becoming an expert in. As part of the privilege of being named the AIChE Management Division award winner, the winner will give a presentation highlighting several of these beyond the technology learnings and failures in hopes that they can be leveraged to help the audience along their leadership journey.

284c Failure: A Catalyst for Success

Daniel Hickman

Technical success in industrial research does not guarantee commercially successful implementation. The potential for research and development to yield a good return on investment requires effective management. Shawn Feist has modeled such leadership in his remarkable career. As Shawn and others have learned, even the best managers and leaders experience failure. A distinguishing feature of those individuals is their ability to harvest lessons from past failures to increase the productivity of future opportunities. This talk will celebrate a few small failures in Dow's research and development function, including one or more examples involving Shawn before he ascended to a management role. By sharing some lessons learned from those failures, we hope to equip others to avoid repeating similar mistakes.

284d Letting Go of Your Dreams: Managing Your Emotions and the Realities of Pivoting or Shutting Down Your Startup.

Nemoy Rau

The mass extinction of startups has begun. The pace of startups shutting down is increasing with fundraising difficulties, undefined business models, and capital being more expensive. Chemical Engineers generally sought deep science related technologies rather than mobile applications causing longer timelines for typical venture capital investors. Your core team members might move on when cash runway is limited, further causing your startup to fail. As a leader, it is critical to navigate the changing technologies and company pain points that might disrupt your company, losing team members, and missing milestones all causing fundraising problems. This two-part talk will focus on: I) the resiliency and convictions of the entrepreneur to believe in his company with market changes and II) managing your emotions, expectations, and lessons learned from shutting down your startup.

284e Don't You Dare Settle for Fine: Cracking the Code of How Dual-Career Couples Thrive in Work and Life.

Rebekah K. Feist

In the U.S. it is estimated that 63% of men and 81% of women are in dual-career relationships, meaning both partners work outside the home.[1] Many of those couples are trying to achieve all they can in their careers, and each has ambitious professional goals. But when two people are dreaming of big careers, a meaningful long-term, successful relationship, and perhaps even a family, it leads to complex challenges that few of us really appreciate. Dual-career couples are constantly grappling with choices to prioritize family, careers, and sometimes one career over the other. Choosing wisely, especially when you don't have all of the data, or time to weigh out the options, can be daunting. What might appear to be the best option today, might also lead to consequences down the road. So how do you know what to select? Understanding what really makes you both happy, how you will measure success in your life, ensuring the risks that you take are really worth the rewards that they could deliver, and making intentional choices every day, big and small, are some of the ways you can support your partner, and together, thrive as a dual-career couple.

89d Fueling the Energy Transition – Lessons Learned from a Renewable Energy Startup

Derek Vardon

Alder Renewables is a dynamic startup with a mission to transform the refinery industry by replacing fossil-based products to rebalance our relationship with the planet. Our primary focus is on scaling our foundational Alder Renewable Crude (ARC) oil platform within existing refinery infrastructure. As the Chief Technology Officer, I have gained invaluable insights during our journey towards deploying ARC, particularly for hard-to-decarbonize sectors like sustainable aviation fuel (SAF).

In this talk, I will share the lessons we've learned from both successes and setbacks on our path to commercializing ARC. These lessons revolve around three key themes:

- Developing strategic collaborations with public and private entities: In our case, we have formed deeply integrated partnerships with Honeywell UOP and the Department of Energy's National Renewable Energy Lab (NREL). These relationships have allowed us to advance our technology, rapidly establish credibility, and navigate complex technical and regulatory landscapes.

- Addressing the pain points in the refinery and aviation fuel communities with ARC: Understanding the core value-drivers of ARC has been instrumental in shaping our technology development strategy. As we have finite resources and timelines, it has been essential to focus on areas that bring the most significant impact to these industries.

- Fostering a team culture of innovation and adaptability: Encountering unforeseen obstacles is inevitable when advancing a new technology. To navigate these challenges, our attitude and actions have been equally critical in maintaining a resilient mindset and steadfast pursuit of our mission to scale and deploy ARC.

89e Scale-up of Purecycle Purification Process

Michael Weber

PureCycle uses a patented, ground-breaking recycling process to separate virtually all color, odor, and contaminants from polypropylene waste feedstock to transform it into Ultra-Pure Recycled (UPR) resin.

This presentation will outline the development of the PureCycle process from the initial concept to commercial reality. A benchtop pressure cell was used to develop the fundamental polymer solubility data that supports the overall design concept. The main unit operations for the process were developed using a lab-scale batch system. Successful operation of the pilot scale system demonstrated that the purification process can be executed continuously. The commercial scale implementation uses a modular design that will enable rapid expansion of the process to multiple sites. The key challenges and successes at each stage of development will be discussed.

At every scale, the purification process has been demonstrated to produce recycled plastic with a color and clarity equivalent to virgin material. This talk will also introduce the next steps for the growth of PureCycle. Future areas of research and optimization will also be discussed.

89a Trials and Tribulations in Entrepreneurship for a Chemical Engineer

Andrew Jones

Chemical Engineers are taught that mastering mass and energy balances, and technical breakthroughs are the keys to building a successful venture. But how many successful businesses are based on breakthrough technology and what does successful execution look like? In this talk, Dr. Andrew Jones will discuss what he has learned over the last nine years through the ups and downs of running two technology-centric innovation companies. Dr. Jones started his first company in 2014 focused on making chemical detectors and add-ons for gas and liquid chromatography. He started his second company in 2021 focused on removing carbon dioxide from the atmosphere and storing it permanently underground as a solid carbon 'charcoal'.

89b Gambling on Innovation with Learning

Darrell Velegol

To maximize your median return (i) and minimize your probability of "going bust" in innovation, you need to invest the proper amount in your innovation projects. A previous article (*IECR*, **60**, 7689, 2021) showed that the Kelly gambling (or investing) strategy helps to make the optimal decisions. To use the Kelly strategy, you must know or estimate the anticipated win ratio (b) of an innovation project and its estimated probability (p) of success. But both b and p will likely increase when you bet a fraction (f) of your innovation budget on a project, due to learning that increases both b and p (*IECR*, **61**, 18457, 2022). This article analyzes how the optimal f changes when learning takes place. I use simple linear approximations $p = p0 + \pi f$ and $b = b0 + f$. As π and/or increases, the optimal Kelly Criterion value fKC increases, often significantly, calling for a higher resource investment, and also giving a dramatic increase in growth rate (i). It is proposed that organizations measure π, , and k, where k is the rate of attempts made per time, and that a primary purpose of an innovating organization is to increase them quickly, both to maximize ROI, and to minimize the probability of "going bust". This talk gives an important "algorithm for innovation".

89c Our Journey in Commercialising Scalable, Economical and Sustainable Porous-Silicon for Li-Ion Battery Anodes

Siddharth Patwardhan

We have discovered a scalable, ultra-low temperature and safer method for producing high-value and stable porous silicon for battery anodes. It overcomes many issues of existing methods such as high production costs, problematic scale-up, unsafe manufacturing process and low-performance products. Currently, we are undertaking steps towards commercialising this technology and, as such, performing (i) process development, (ii) product testing, and (iii) market discovery. In this presentation, we will share our efforts in the commercialisation journey. Special emphasis will be on market discovery side, starting from understanding the commercial opportunity, exploring commercialisation options and developing mechanisms to spin-out.

121a Reducing Methane Emissions and Creating Value from Stranded Associated Gas

Paul Yelvington

Stranded associated gas, produced as a byproduct of tight oil production, is an underutilized feedstock that is commonly flared, vented, or reinjected. Improved utilization of stranded gas faces two primary technical/economic challenges. First, the gas is inherently distributed geographically and the supply varies temporally, which requires rethinking the characteristics of the chemical plant used to process the gas. Second, methane—the primary constituent of associated gas—is relatively inert and requires activation through a synthesis gas intermediate prior to conversion to useful products. This methane activation step requires reforming chemistry which adds to the cost and complexity of the plant. This presentation explores the potential of piston engines as non-catalytic, partial-oxidation reformers, providing an enabling technology for distributed production of chemicals from stranded gas resources. The motivation for using an engine in this process is to take advantage of the modular, scalable, fast-responding characteristics of engines, leveraging this highly developed and mass-manufactured device for purposes beyond power generation or transportation. This presentation provides a perspective on technical, economic, regulatory, and environmental aspects of engine reforming for production of low-carbon fuels and chemicals from stranded gas.

121b Sustainability in Capital Investments

James Fisher

Amgen embarked on our latest and most aggressive Sustainability Plan which we call Amgen's 2027 Ambition. This plan will deliver carbon neutrality, a 40% reduction in water (from a 2019 baseline), and a 75% reduction in waste disposed (from a 2019 baseline) by the end of year 2027. Amgen's plan is backed by the Board of Directors, will require over $200 million investment, and relies heavily on enterprise-wide efforts for success. In addition to sustainability driven projects, Amgen has improved our Engineering Project Delivery process to assess all Investment Lifecycle Management capital projects for both sustainability impact and opportunity to introduce sustainability related scope. This ensures all projects justified by other drivers, such as reliability or capacity, consider and support Amgen's commitment to sustainability.

121d 3M Sustainability – Applying Science to Improve Lives

Derek Dehn

This presentation will lay out 3M's sustainability goals as they relate to circular, climate, and community. Specific examples of select current projects will be provided to help demonstrate how 3M's innovation is working to improve lives, solve customer problems, and better the environment for current and future generations.

121e Unleashing Hydrogen's Power

Jason McMullan

Energy is essential for human development and improving living standards around the world. Society faces a challenge to provide reliable and affordable energy to meet the needs of growing population, while reducing environmental impacts, including the risks of climate change. Progress on society's energy and climate objectives requires practical approaches, new technology solutions, and collaborations and partnerships that enable human development and economic progress. ExxonMobil established its Low Carbon Solutions business in early 2021 focusing on carbon capture and storage (CCS), hydrogen and biofuels, leveraging our core capabilities in technology, scale, integration, and people. This presentation will provide an overview of energy fundamentals, ExxonMobil's strategy, and innovative approaches to technology development in the Hydrogen area.

190c DEI Programs at Tickle College of Engineering, University of Tennessee

Jeremy Kirk

Diversity, equity and inclusion is the hallmark of the human experience. To be able to live, worship, cultivate, speak, work, in the company of people from various demographic backgrounds is a privilege. In the Tickle College of Engineering, we strive to ensure ALL stakeholders know that through similarities and differences, respect of people and culture is paramount. Closing in on 50 years of celebrating engineering diversity programs in the College, we are working tirelessly to create sustainable initiatives that mirror the success of our program through the years. We align our priorities with that of the University and College leadership to measure outcomes according to our stakeholders. We are shaping our students, faculty and staff into agents of great intellectual capabilities and understanding that bring all walks of life to the table. Our legacy is strong and incorporating DEI into all of our daily processes continues to make us stronger for the future of the College.

190d Partnerships Drive UCOR Workplace Recruitment and Retention

Tommy Morgan

UCOR, an Amentum-led partnership with Jacobs, is strongly committed to developing the current and future workforce to accomplish the hazardous multi-decade environmental cleanup of land and aging contaminated nuclear facilities on the Department of Energy's Oak Ridge (TN) Reservation. Successful completion demands well-trained and highly-skilled workers experienced in their discipline or craft and dedicated to safe operations. This presentation will focus on the recruitment, training and retention of the UCOR workforce through multiple avenues, including UCOR in-house training, internships and partnerships with trade unions, regional higher educational institutions, HBCUs, and regional technical schools. Examples include establishment of UCOR's university consortium, promoting participation in STEM-related fields, and increasing interest by minority students in science and engineering careers.

190e Siemens' Training Initiative in Developing Modelling Skills for Chemical Engineering Graduates

Udit Gupta

Digitalization and analytics have become an intrinsic part of industry, which subsequently, has resulted in an increase in the demand for highly skilled professionals. Additionally, novel sustainable technologies require the development of process models that are different from conventional unit operations. These higher requirements for better trained students have put pressure on Chemical Engineering faculties to revisit their academic programmes, in order to ensure that their graduate engineers have the necessary set of skills to face the many challenges in the modern world.

There is clear evidence of a gap between what industry requires and expects and what academia can provide. Siemens has been active in bridging this gap by collaborating with universities and facilitating teaching of modelling and simulation in academic programmes. Siemens partners with universities around the world by providing technical content and material that covers different topics such as mass and energy balances, separation methods, thermodynamics, flowsheeting, optimization, among others. This material is modular and flexible so lecturers can incorporate modelling topics throughout the curriculum without taking time away from fundamentals or replacing any other content. Siemens has seen different levels of success with this approach, showing that this effort is a viable and practical way of teaching modelling tools in Chemical Engineering programmes.

190f Bridge to Bio: Workforce Development in Biopharmaceutical Manufacturing

George Prpich

Bridge to Bio is an immersive workforce development program aimed at increasing the diversity of new engineering talent entering the biomanufacturing sector through the Virginia Community College network. A collaboration between the University of Virginia (UVA), Merck, and Pfizer, the program integrates three evidence-based pillars of change: 1) increasing representation, 2) fostering community and belonging, and 3) working to sustain change by building effective industrial relationships. Bridge to Bio began January 2023, with a 5-day immersion program. Fourteen students from across the Virginia Community College network participated in programming that provided insight about how life-saving medicines are manufactured and how to access university engineering programs. Participants also enjoyed a 1-day visit to Merck's Elkton, VA facility where they received a hands-on experience performing day-to-day manufacturing activities. Keys to program success thus far have been network development, effective messaging, and industry engagement. This presentation will provide an update and a description of the next steps, as well as allow us to share our perspectives on the value of meaningful and effective working relationships with pipeline partners.

330a From Chemical Engineering to Intellectual Property Law

Paul Townsend

My name is Paul Townsend, and I am an intellectual property attorney at the law firm Finnegan, Henderson, Farabow, Garrettt & Dunner, LLP, in Washington, DC. I didn't start as an attorney though; I originally went to school in Canada, Queen's University, for chemical engineering. When people here this, the first question I always get asked is whether I always planned to go to law school after becoming a chemical engineer. And the answer to that is no. I never even considered law school until I was almost done with my engineering degree. So how did I get here? It starts with my work experiences as a chemical engineering student. When I was studying at Queen's University, I worked in different jobs all three summers. Each job was interesting, but there was never anything that made me really excited about working for those companies. When I realized that those types of jobs were what I would likely be working in, I decided to make a change. That's when I first learned about patent law, a job where I could apply legal reasoning and my technical background.

After getting my law degree, I got a job at Finnegan Henderson doing intellectual property law. I've been working at the law firm for about 6 years now, so I'm still fairly early in my career. Nonetheless, I've had a lot of great experiences in those 6 years. My focus over these past years has been in two areas of intellectual property: patents and trade secrets. I imagine most of you know about patents. Those are the government granted protections on something that you invent. Trade secrets are a little less clear, but those always involve some secret that you or your company obtains an economic benefit from. My work at Finnegan has evolved over the 6 years I've been here. I've had experience meeting with inventors and drafting patents to protect their inventions, implementing intellectual property policies at various companies, providing all sorts of legal advice regarding intellectual property, and litigating all sides of intellectual property disputes in court.

During my presentation, I'll give you a more detailed description of how I got where I am now. I'll then provide more detail on what exactly I do at Finnegan and how my role has changed over the years.

330b From a PhD in Cheme to Science Policy and a Government Career

Ashley Pennington

Dr. Pennington will present an oral history of choices made, and the potential alternatives not made, which lead her from the laboratory to the Federal government for a career in science policy where she leverages her PhD in Chemical and Biochemical Engineering to enhance governmental leaders and policy maker's scientific understanding. Of the items covered, will be the decision to obtain a Masters and subsequent PhD, the decision of where to serve as a Postdoctoral research associate, and the conversion from laboratory work to science policy.

After obtaining her PhD at Rutgers University, Dr. Pennington became an NRC Postdoctoral Associate at the Naval Research Laboratory where she worked on research projects with DTRA and naval interest. From the laboratory-based postdoctoral fellowship Dr. Pennington was accepted to the non-laboratory-based AAAS Science and Technology Policy fellowship where she worked with DOE's Federal Energy Management Program to leverage her engineering background to enhance public-private partnerships pertaining to the integration of fleet electrification, utility partnerships, and climate resilience. Dr. Pennington is now the Stakeholder Engagement Division's Chemical Engineer at the Department of Homeland Security's Cybersecurity and Infrastructure Security Agency (CISA). In her role, Dr. Pennington provides technical expertise and advice relating to the supply chain, physical and cybersecurity, and climate resilience of the chemical sector, as CISA is the Sector Risk Management Agency (SRMA) for the nation's chemical sector.

330c The Lifeline of Chemical Researcher in a Business Environment

Emilio Nehme

Imagine traveling back in time to our first day at work and asking ourselves the one million dollar question "should have we performed differently to satisfy, our colleagues, our direct reporting line, and importantly ourselves?" In fact it's not only one but too many breaking down structure that comes to our mind, enabling us to draw a lifeline of our career path with sinusoidal achievements and hiccups or pitfalls events whether internally at your work environment or externally within your personal life but in both case having huge impact on your performance. From my own perspective, drawing this lifeline was not an easy task when one of the guru minds in management asked us in leadership training to do so. How to include 25 years of career history in a trend correlated with our personal life. Starting from my first day when I joined the company as quality control Laboratory technician with only my first year major in chemistry, then a double major master degrees in environmental chemistry and biochemistry that helped me a lot to join the sustainability department and understand the different management systems applied at the company from a strategically point of view and promoted to be the Environmental specialist until currently holding the position of Business Partner Environment which widen a lot the scope of my responsibility all over the plant. Nevertheless this line will continue and the growth of our knowledge never been satisfied to fill what so called our "think tank" and being in line with the world evolvement around us, which encourages me to continue my doctorate research in management hopefully that will open for me new chapters in this book of career path.

330d A Career in Managment for a Ph.D. Chemical Engineering Generalist

Joseph Cramer

The author will use his own career path to illustrate how a sound educational foundation in chemical engineering enables one to function well in a variety of different career roles. Dr. Cramer initially started out along a career pathway that typically leads one to a career in R&D but a difficult employment market led to a first industry position with an engineering company. Discovering that he enjoyed the challenges and excitement of working on specific assignments and the interactions with clients and project team members gradually induced him into project and later technical department management. After more than 20 years of working in a consulting engineering environment the employment market once again made him change directions. This time he landed a position working for a technically oriented not-for-profit organization, none other than AIChE. The work was even more management and people oriented than his industry experience but now instead of having a more direct form of authority over a modest number of direct supports he had to develop an ability to influence large numbers of volunteer professionals. As the staff member responsible for guiding the development of much of the content for AIChE technical meetings he continued in this capacity for 18+ years until he retired and returned to being a volunteer with both AIChE and the National Council for Examiners of Engineering and Surveying. Throughout his entire career path, the principles he learned as both an undergrad and grad student in Chemical Engineering served him well and allowed him to adapt as required.

AUTHOR INDEX